Make it E[asy]

Age 5-6

Maths

Contents

Learning Activities

Quick Tests

Answers

Paul Broadbent and Peter Patilla

Numbers to 10

Look at these numbers and say them out loud.

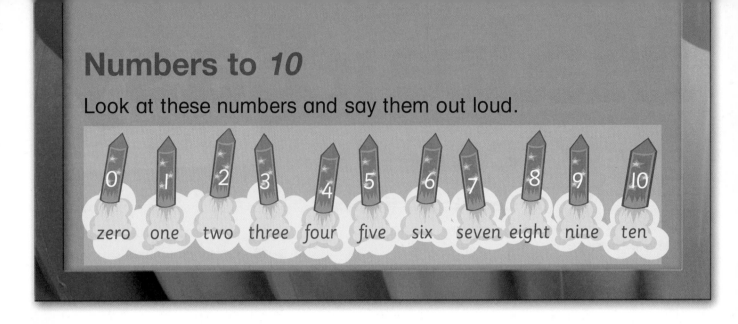

I **Draw over the numbers. Join each one to its matching picture.**

1 2 3 4 5 6 7 8 9 10

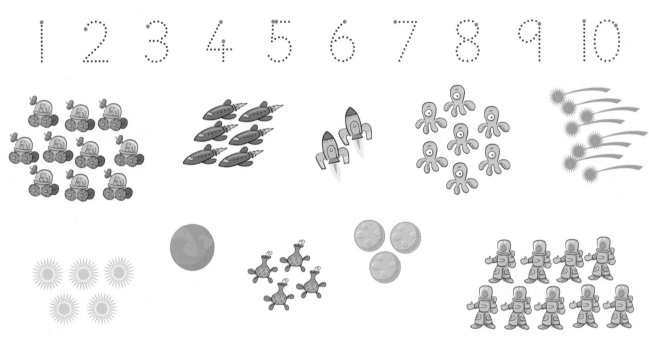

II **Write the number on each planet to match each word.**

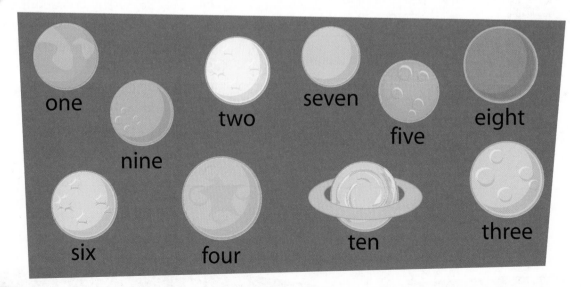

one

two

seven

eight

nine

five

six

four

ten

three

2

Counting

Count the shells and say the numbers out loud.

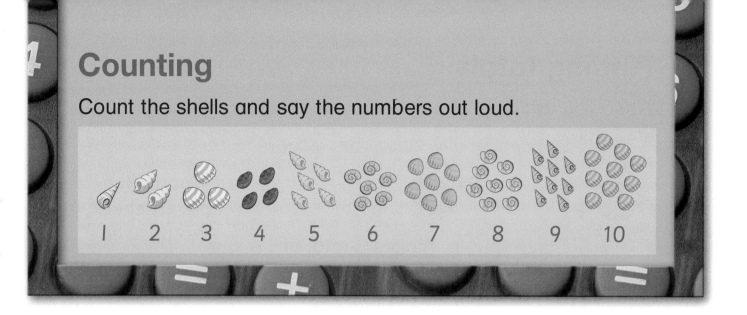

1 2 3 4 5 6 7 8 9 10

I Count the things in each group. Write the number.

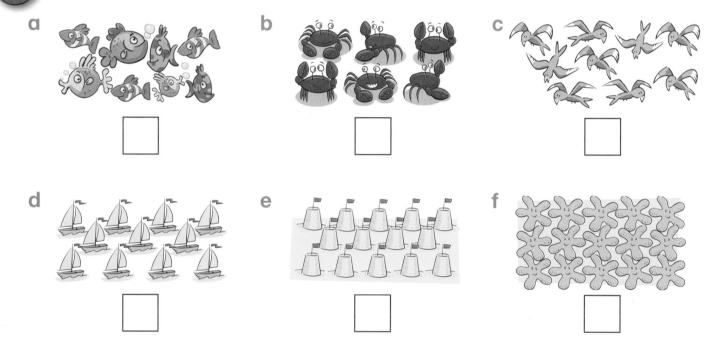

a

b

c

d

e

f

II Draw 10 more fish in the pool.

There are ⬜ fish altogether.

Finding totals

When you **add** sets of objects together, you are finding the total.

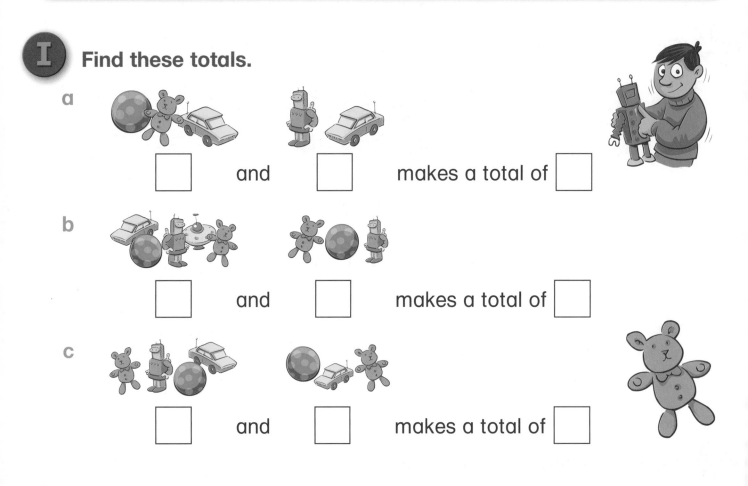

4 and 2 makes a total of 6

Count the cars to check the answer.

I Find these totals.

a

☐ and ☐ makes a total of ☐

b

☐ and ☐ makes a total of ☐

c

☐ and ☐ makes a total of ☐

II Draw extra beans so each set makes a total of 8.

8

2-D shapes

Look around you for these 2-D shapes. Try to remember their names.

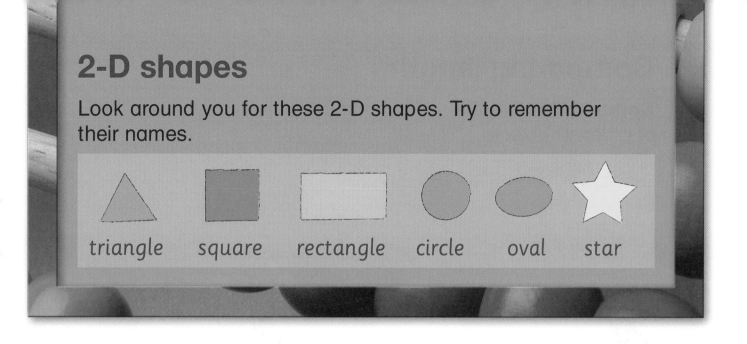

triangle   square   rectangle   circle   oval   star

I   Colour the shapes to match the code. Count the number of each shape.

Number of shapes

II   Draw lines to join each shape to its name.

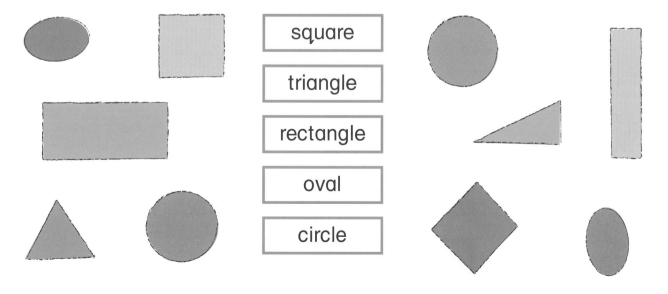

square

triangle

rectangle

oval

circle

Comparing lengths

Compare the lengths of different objects.

Use pennies to measure the lengths.

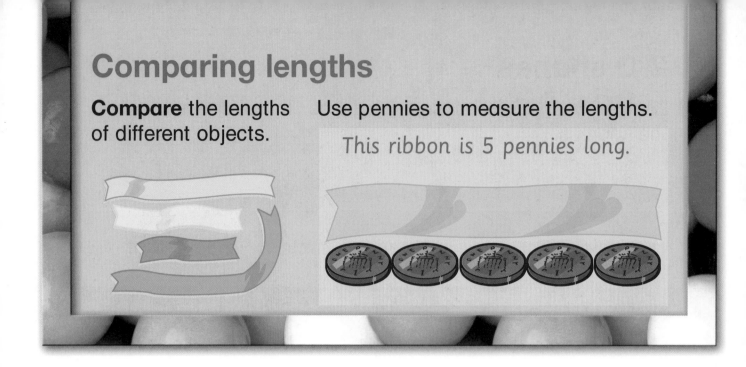

This ribbon is 5 pennies long.

I Look at these objects.

a Circle the longest in each group.

b Circle the shortest in each group.

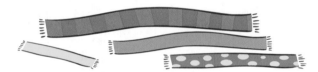

II Measure these lines using pennies.

a ☐ pennies

b ☐ pennies

c ☐ pennies

d ☐ pennies

e ☐ pennies

O'clock time

When the long minute hand points to 12, it is an **o'clock time**.

On a digital clock, an o'clock time ends with **00**.

This is 4 o'clock.

 Write the time shown on each clock.

a

[ ] o'clock

b

[ ] o'clock

c

[ ] o'clock

d

[ ] o'clock

e

[ ] o'clock

f

[ ] o'clock

g

[ ] o'clock

h

[ ] o'clock

 Here are some activities Jack does each Saturday. Show the time he finishes each of them.

a

start ➜ 1 hour ➜ finish

b

start ➜ 3 hours ➜ finish

c

start ➜ 2 hours ➜ finish

d

start ➜ 1 hour ➜ finish

Numbers to 20

Look at these numbers and say them out loud.

11 eleven	**14** fourteen	**17** seventeen	**20** twenty
12 twelve	**15** fifteen	**18** eighteen	
13 thirteen	**16** sixteen	**19** nineteen	

I Draw over each number, starting at the red dot.

11   12   16   17

14   13   18   19

15   20

II Write the number to match each word.

a 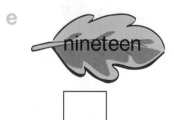 fourteen ☐

b eighteen ☐

c thirteen ☐

d sixteen ☐

e 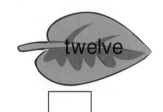 nineteen ☐

f twelve ☐

g eleven ☐

h fifteen ☐

Patterns and sequences

Colours, **shapes** and **lines** can make different patterns and sequences.

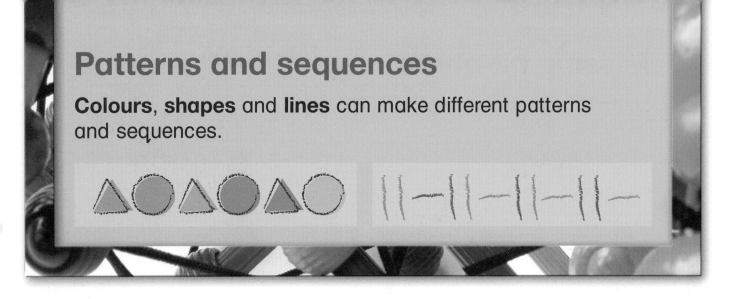

I. Continue drawing each pattern.

a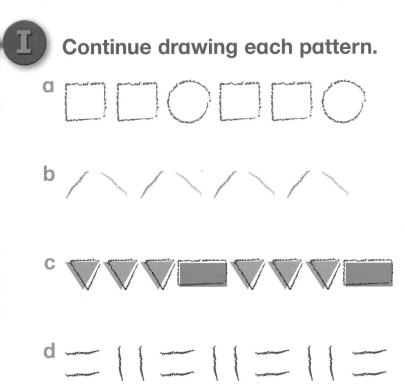

b

c

d

II. Draw over these. Colour to make a pattern.

a

b

Missing numbers

Learn the order of numbers to **20**.

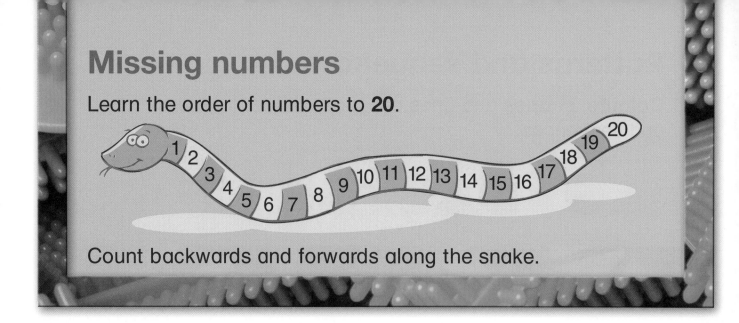

Count backwards and forwards along the snake.

I Write the missing numbers.

a   7   8   ☼   ☼   11   ☼   ☼   14   ☼   ☼

b   ☼   ☼   ☼   ☼   15   16   17   ☼   ☼   20

c   18   17   16   ☼   ☼   13   ☼   ☼   ☼   ☼

d   ☼   ☼   10   9   ☼   ☼   ☼   5   4   ☼

II Write these numbers as words. What is the hidden, shaded number?

12	→
15	→
18	→
13	→
10	→

The shaded number is ⬜.

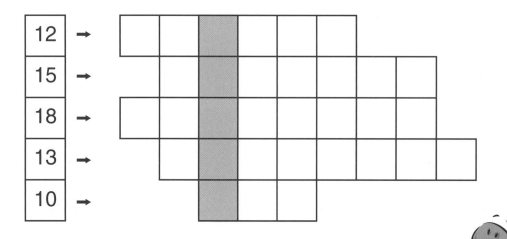

Counting forwards and backwards

Practise counting **forwards** and **backwards** on a number line.

Follow the jumps with your finger.

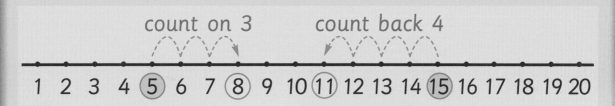

I Write the answers. Use the number line above to help.

a 3 → Count on 2 → ☐

b 7 → Count on 3 → ☐

c 11 → Count on 5 → ☐

d 10 → Count on 4 → ☐

e ☐ ← Count back 2 ← 8

f ☐ ← Count back 4 ← 12

g ☐ ← Count back 3 ← 16

h ☐ ← Count back 5 ← 14

II Draw the jumps to show the counting on or back. Circle the number you finish on.

a Count on 4

c Count back 2

b Count on 3

d Count back 5

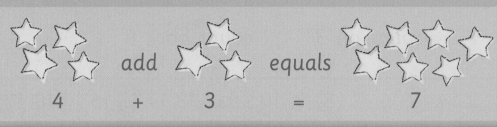

Starting to add

We use the **+** sign to show **adding**.

= is the **equals** sign.

4   +   3   =   7

 Write the numbers for these additions.

a

☐ + ☐ = ☐

c

☐ + ☐ = ☐

b

☐ + ☐ = ☐

d

☐ + ☐ = ☐

 Draw some more spots to help work out the missing numbers.

a

3   +   ☐   = 7

c

4   +   ☐   = 6

b

2   +   ☐   = 5

d

5   +   ☐   = 9

12

3-D shapes

Look around you for these 3-D shapes. Try to remember their names.

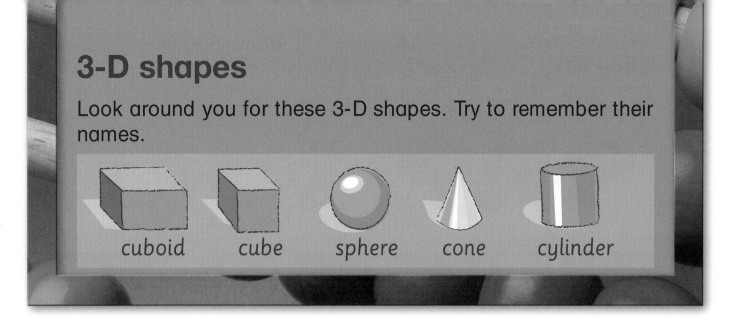

cuboid   cube   sphere   cone   cylinder

I **Draw lines to join each shape to its name.**

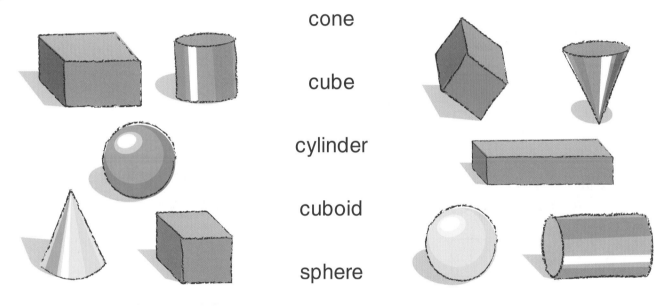

cone

cube

cylinder

cuboid

sphere

II **Colour in the shapes that have all flat faces.**

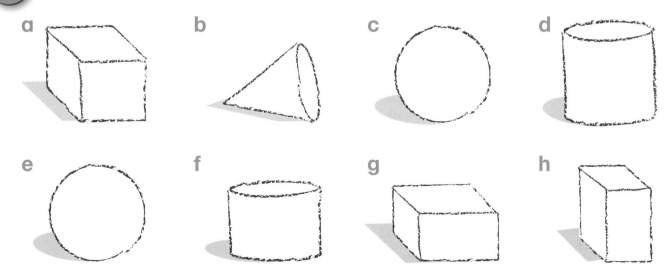

a   b   c   d

e   f   g   h

Starting to take away

We use the – sign when we **subtract** or **take away**.

$$5 - 2 = 3$$

start with 5   take away 2   3 left

I Cross out some buns to help answer these.

a

$$5 \ - \ 4 \ = \ \square$$

c

$$4 \ - \ 2 \ = \ \square$$

b

$$6 \ - \ 2 \ = \ \square$$

d

$$6 \ - \ 3 \ = \ \square$$

II Write out a subtraction for each of these.

a

$$\square \ - \ \square \ = \ \square$$

c

$$\square \ - \ \square \ = \ \square$$

b

$$\square \ - \ \square \ = \ \square$$

d

$$\square \ - \ \square \ = \ \square$$

Recognising coins

Try to learn these **coins**.

| 1p | 2p | 5p | 10p | 20p | 50p |

I Cross out the odd one in each set.

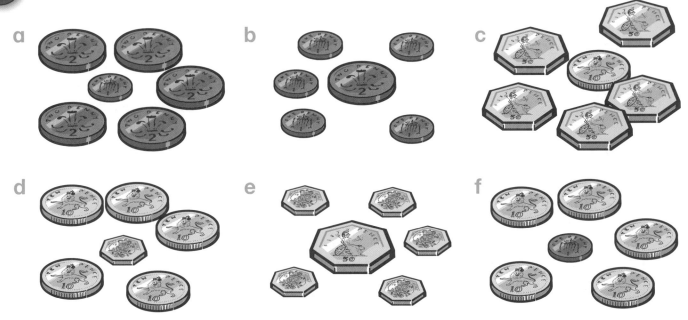

a

b

c

d

e

f

II Write the totals for each of these.

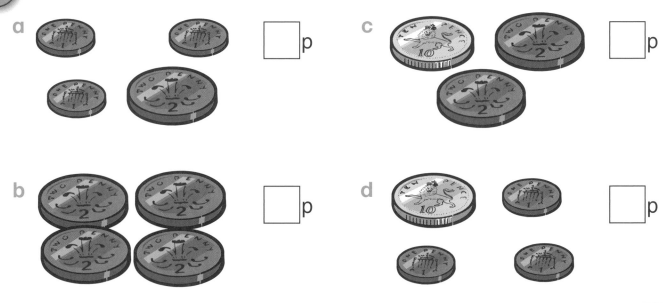

a ☐p

c ☐p

b ☐p

d ☐p

More or less

Look at the way numbers change if you make them **1** or **10** more or less.

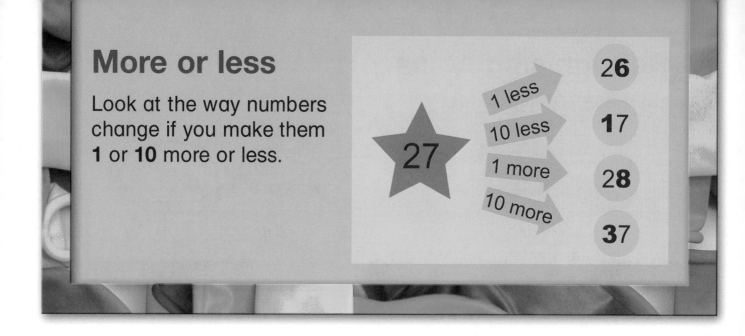

27

1 less → 26
10 less → 17
1 more → 28
10 more → 37

I **Answer these.**

a Add 1 more.

13 → ◇   18 → ✶

26 → ○   31 → □

c Add 10 more.

7 → ◇   19 → ✶

14 → ○   22 → □

b Make 1 less.

19 → ◇   15 → ✶

26 → ○   20 → □

d Make 10 less.

18 → ◇   21 → ✶

32 → ○   29 → □

II **Count how much money is in each purse.**

a Add 1p and write the total amount.

b Add 10p and write the total amount.

 p
 p
 p
 p

Odds and evens

Say these odd and even numbers out loud.

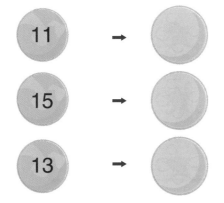 **Try to do these odd and even problems.**

a Tick the shields with an even number of dots.

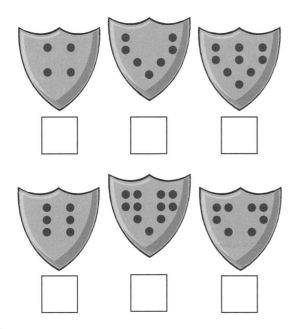

b Colour all the badges with even numbers.

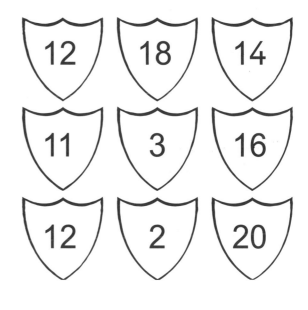

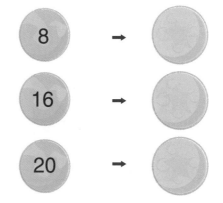 **Now try these problems.**

a Write the next odd numbers.

11 →

15 →

13 →

b Write the next even numbers.

8 →

16 →

20 →

Ordering numbers

A number track will help you to learn the order of numbers.

| 0 | 1 | 2 | 3 | 4 | 5 | 6 | 7 | 8 | 9 | 10 | 11 | 12 | 13 | 14 | 15 | 16 | 17 | 18 | 19 | 20 |

Cover some numbers with your fingers without looking. Use the other numbers to work out which ones you have hidden.

I These numbers have fallen off the washing on each line. Put them back in the correct order.

a

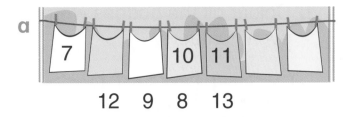

| 7 | | | 10 | 11 | | |

12 9 8 13

c

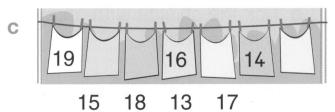

| 19 | | | 16 | | 14 |

15 18 13 17

b

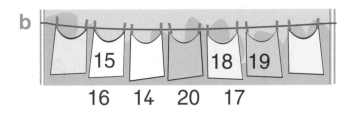

| 15 | | | 18 | 19 | |

16 14 20 17

d

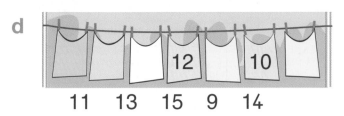

| | | 12 | | 10 |

11 13 15 9 14

II Draw lines to join the price labels to the correct tins. The prices should be in order.

18p   9p   13p   14p

7p   15p   11p   20p

Lowest ⟶ Highest

Teen numbers

Teen numbers are made from a **10** and some **1s**.

Read these numbers aloud.

13	14	15	16	17	18	19
10+3	10+4	10+5	10+6	10+7	10+8	10+9

11 and 12 are also made from a 10 and some 1s.

I Write the missing numbers.

a fifteen   ➡ $\boxed{10}$ + $\boxed{}$ f twelve   ➡ $\boxed{}$ + $\boxed{}$

b eighteen ➡ $\boxed{}$ + $\boxed{}$ g fourteen ➡ $\boxed{}$ + $\boxed{}$

c eleven   ➡ $\boxed{}$ + $\boxed{}$ h thirteen ➡ $\boxed{}$ + $\boxed{}$

d sixteen   ➡ $\boxed{}$ + $\boxed{}$ i seventeen ➡ $\boxed{}$ + $\boxed{}$

e nineteen ➡ $\boxed{}$ + $\boxed{}$ j twenty   ➡ $\boxed{}$ + $\boxed{}$

II Write these answers in words.

a 10 + 3 ➡

b 10 + 1 ➡

c 10 + 9 ➡

d 10 + 4 ➡

e 10 + 2 ➡

f 10 + 6 ➡

g 10 + 7 ➡

h 10 + 8 ➡

Comparing numbers

When you compare numbers you can use a number line to work out which number is bigger and which is smaller.

```
•    •    •    •    •
12   13   14   15   16
```

15 is bigger than 12.

I Look at these pairs of numbers.

a Colour the bigger number red.

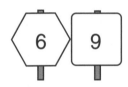
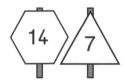

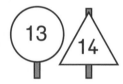

b Colour the smaller number blue.

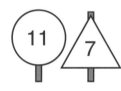
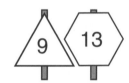

II Write the missing numbers between these pairs of numbers.

a

d

b

e

c

f

Ordinal numbers

1st, 2nd, 3rd ... are called ordinal numbers.

They show the **order of things**.

first	second	third	fourth	fifth	sixth	seventh	eighth
1st	2nd	3rd	4th	5th	6th	7th	8th

I **Look at the order of the letters in the alphabet.**

A B C D E F G H I J K L M N O P Q R S T U V W X Y Z

Complete these.

The 1st letter is ☐.   F is the ☐ letter.

The 5th letter is ☐.   C is the ☐ letter.

The 7th letter is ☐.   L is the ☐ letter.

The 4th letter is ☐.   P is the ☐ letter.

The last letter is ☐.   B is the ☐ letter.

II **Use the alphabet order to work out these word puzzles.**

a 13th   1st   20th   8th   19th

☐M☐ ☐ ☐ ☐ ☐

b 13th   1st   7th   9th   3rd

☐M☐ ☐ ☐ ☐ ☐

Now try making up your own puzzles.

Days of the week

Try to learn the **order** of the days of the week.

Sunday Saturday Monday Friday Tuesday Thursday Wednesday

I How well have you learnt the days of the week?

a Draw lines to join each day to the one that follows it.

Friday
Wednesday
Monday
Sunday
Tuesday
Thursday
Saturday

b Now join each day to the one that comes before it.

Saturday
Monday
Thursday
Friday
Sunday
Tuesday
Wednesday

II Fill in the gaps to show what day it is.

W e __ __ __ __ d a y

S __ __ u r __ __ y

F __ __ __ a y

__ o n d __ __

S u __ __ a y

__ __ u r s d __ __

T __ e __ d __ y

MONDAY
TUESDAY
WEDNESDAY
THURSDAY
FRIDAY
SATURDAY
SUNDAY

Money

You need to know what these **coins** are worth.

1p 2p 5p 10p 20p 50p

p means pence. There are 50 pence or pennies in 50p.

I Add up the value of these coins to find the total.

a [ ]p

c [ ]p

b [ ]p

d [ ]p

II Draw the coins you would use to buy these sweets.

a 7p

b 16p

c 11p

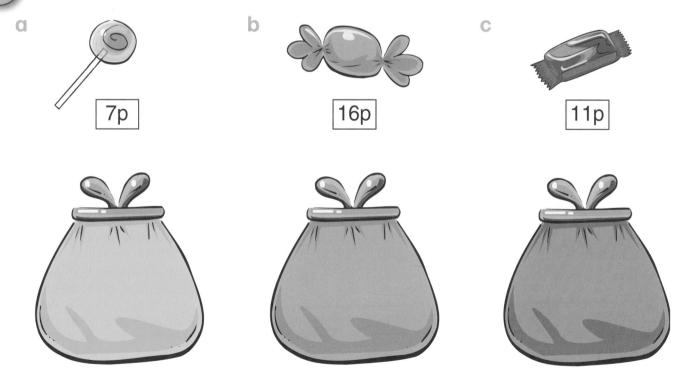

Adding

Use a **number track** like this to help you add.

$$2 + 3 = 5$$

+3

0 1 2 3 4 5 6 7 8 9 10

I Use the number track to help you add these.

a 3 + 4 = ☐   f 2 + 5 = ☐   k 4 + 5 = ☐

b 1 + 3 = ☐   g 4 + 4 = ☐   l 6 + 3 = ☐

c 4 + 2 = ☐   h 3 + 2 = ☐   m 5 + 1 = ☐

d 0 + 2 = ☐   i 0 + 5 = ☐   n 8 + 2 = ☐

e 3 + 3 = ☐   j 5 + 3 = ☐   o 3 + 7 = ☐

II Write the numbers coming out of the machines.

a ☐ 3   ☐   e ☐ 2   ☐

b ☐ 5   ☐   f ☐ 5   ☐

+ 4   + 5

c ☐ 2   ☐   g ☐ 4   ☐

OUT   OUT

d ☐ 6   ☐   h ☐ 1   ☐

Taking away

You can count back along a number line to help you **subtract** or **take away**.

$$8 - 3 = 5$$

-3

4   5   6   7   8   9   10

I Use the number lines to help you complete these.

a   5   6   7   8   9   10

9 – 4 = ☐

b   1   2   3   4   5   6   7

7 – ☐ = 3

c   2   3   4   5   6   7   8

8 – ☐ = 4

d   5   6   7   8   9   10

10 – ☐ = 7

e   2   3   4   5   6   7   8   9

8 – 5 = ☐

f   1   2   3   4   5   6   7

5 – ☐ = 3

II Colour the tyre which gives a different answer to the others in the pile.

a
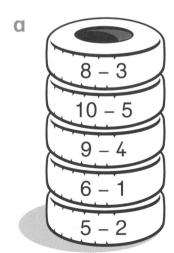

8 – 3
10 – 5
9 – 4
6 – 1
5 – 2

b
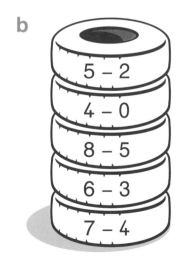

5 – 2
4 – 0
8 – 5
6 – 3
7 – 4

c

8 – 6
6 – 5
4 – 2
5 – 3
9 – 7

Measuring

Mass and capacity are 2 ways of measuring.

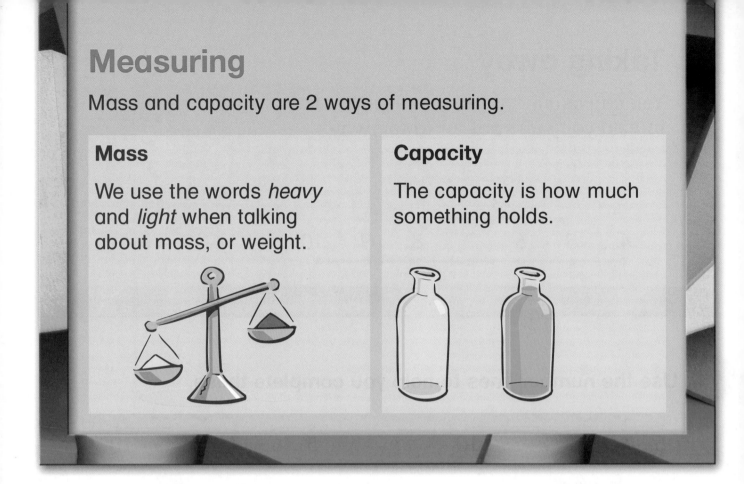

Mass

We use the words *heavy* and *light* when talking about mass, or weight.

Capacity

The capacity is how much something holds.

I Which objects are heavy and which are light? Circle the lightest ones.

a   b   c

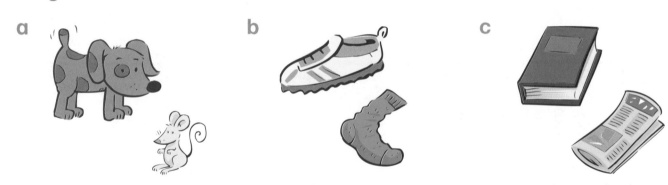

II Look at these containers. Draw lines to join them in order. Start with the smallest capacity.

Counting patterns

Use this number grid to help you with **counting patterns**.

1	2	3	4	5	6	7	8	9	10
11	12	13	14	15	16	17	18	19	20
21	22	23	24	25	26	27	28	29	30
31	32	33	34	35	36	37	38	39	40

I **Look at the numbers on the track.**

Colour the number 2 red.

Miss out number 3 and colour number 4 red.

Miss out number 5 and colour number 6 red.

Continue colouring this pattern.

1 2 3 4 5 6 7 8 9 10 11 12 13 14 15 16 17 18 19 20

a The red numbers are called _____ numbers.

b The other numbers are called _____ numbers.

II **Continue these counting patterns.**

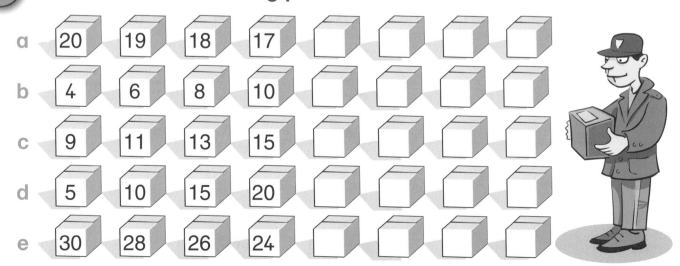

a 20 19 18 17

b 4 6 8 10

c 9 11 13 15

d 5 10 15 20

e 30 28 26 24

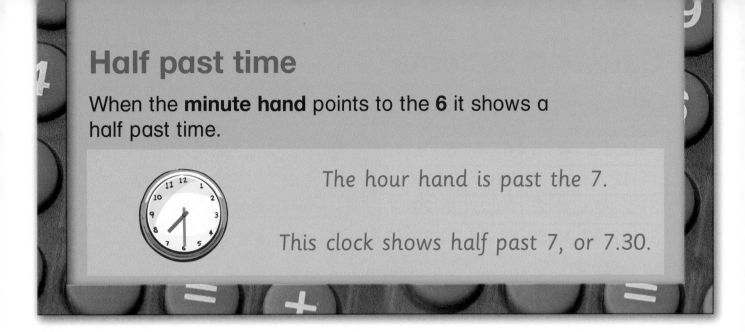

Half past time

When the **minute hand** points to the **6** it shows a half past time.

The hour hand is past the 7.

This clock shows half past 7, or 7.30.

 Write out these times in the boxes.

a

half past ☐

c

half past ☐

e

half past ☐

b

half past ☐

d

half past ☐

f

half past ☐

 Draw lines to join the clocks that show the same time.

a

b

c

 4:30

 9:30

 1:30

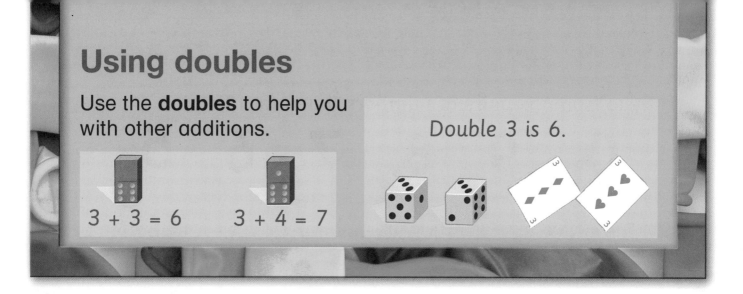

Using doubles

Use the **doubles** to help you with other additions.

3 + 3 = 6   3 + 4 = 7

Double 3 is 6.

 Write the answers in the boxes.

a  3 + 3 = ☐

b 2 + 2 = ☐

c 5 + 5 = ☐

d 4 + 4 = ☐

e 1 + 1 = ☐

f  2 + 3 = ☐

g 4 + 5 = ☐

h 1 + 2 = ☐

i 3 + 4 = ☐

j 5 + 6 = ☐

II Draw more dots on each domino to match the total in the box below.

a    b    c    d    e

| 9 | 7 | 5 | 11 | 12 |

Totalling 10

Try to learn the **pairs of numbers** that total 10.

$$0_{10} \quad 9_1 \quad 8_2 \quad 7_3 \quad 6_4 \quad 5_5$$

The order in which you add numbers does not matter.
8 + 2 and 2 + 8 both total 10.

I Write the missing numbers.

a 3 + ☐ = 10

b 8 + ☐ = 10

c 4 + ☐ = 10

d 5 + ☐ = 10

e ☐ + 0 = 10

f ☐ + 9 = 10

g ☐ + 8 = 10

h ☐ + 7 = 10

i 1 + ☐ = 10

j ☐ + 4 = 10

k 10 + ☐ = 10

l ☐ + 5 = 10

II Corner numbers add up to 10. Write the missing corner numbers for each of these.

a

4

10

1 — ○

c

6

10

2 — ○

e

3

10

○ — 1

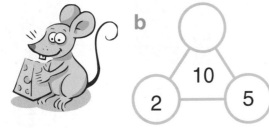

b

○

10

2 — 5

d

1

10

○ — 5

Finding the difference

A **number track** is very useful for finding the difference between 2 numbers.

Count on from 4 to 7, counting the jumps.

The difference between 4 and 7 is 3.

I Draw the jumps to find the difference between these pairs of numbers in red.

a

The difference is ☐.

c

The difference is ☐.

b

The difference is ☐.

d

The difference is ☐.

II Draw lines to join the pairs of numbers with a difference of 4.

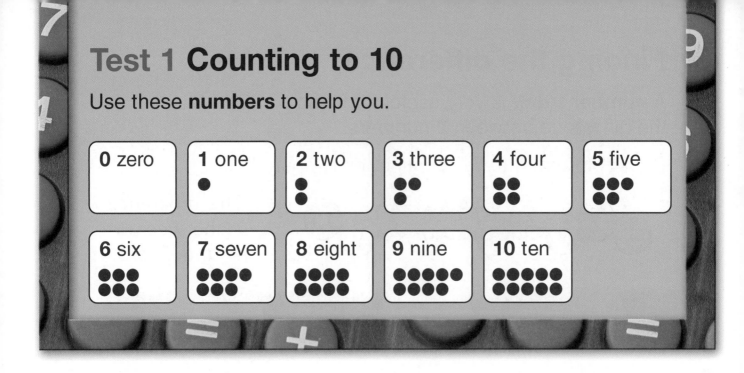

Test 1 Counting to 10

Use these **numbers** to help you.

0 zero	**1** one	**2** two	**3** three	**4** four	**5** five
6 six	**7** seven	**8** eight	**9** nine	**10** ten

Count how many marbles. Write the total in the box.

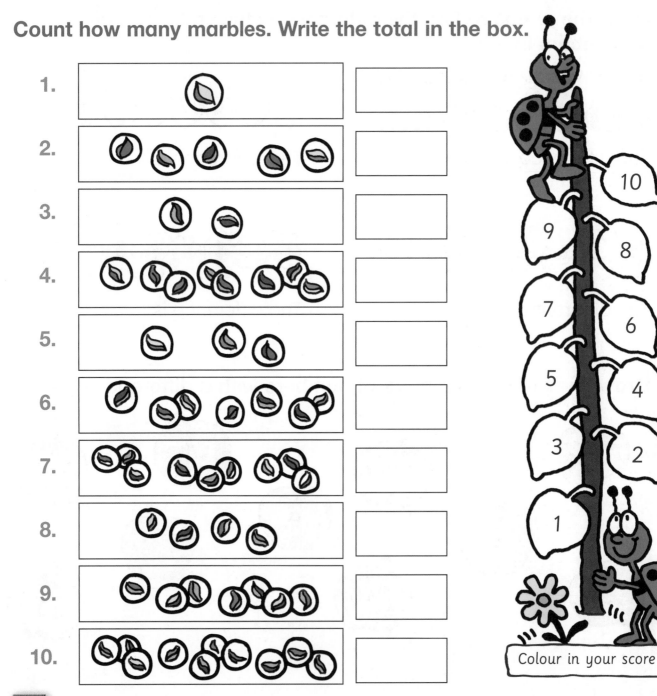

1.

2.

3.

4.

5.

6.

7.

8.

9.

10.

Colour in your score

Test 2 Reading and writing numbers to 20

Use these **numbers** and **words** to help you.

0 zero	**1** one	**2** two	**3** three	**4** four	**5** five
6 six	**7** seven	**8** eight	**9** nine	**10** ten	
11 eleven	**12** twelve	**13** thirteen	**14** fourteen	**15** fifteen	
16 sixteen	**17** seventeen	**18** eighteen	**19** nineteen	**20** twenty	

Write the number.

1. six

2. eight

3. twelve

4. fifteen

5. twenty

Write the word.

6. 3

7. 9

8. 11

9. 17

10. 19

Colour in your score

33

Test 3 Addition to 10

Number lines help you to **add**.

$$3 + 4 = 7$$

Write the answers to these sums.

1. 6 + 3 =

2. 2 + 5 =

3. 7 + 2 =

4. 4 + 4 =

5. 5 + 4 =

6. 1 + 7 =

7. 5 + 3 =

8. 6 + 4 =

9. 3 + 7 =

10. 8 + 1 =

Colour in your score

34

Test 4 Measures: length

shortest   tallest

Tick the shortest.

1.

6.

2.

7.

3.

8.

4.

9.

5.

10.

10
9
8
7
6
5
4
3
2
1

Colour in your score

35

Test 5 2-D shapes

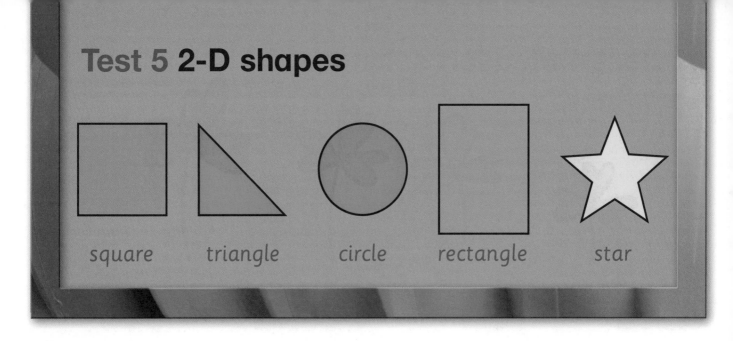

square	triangle	circle	rectangle	star

Name each shape.

Finish drawing each shape.

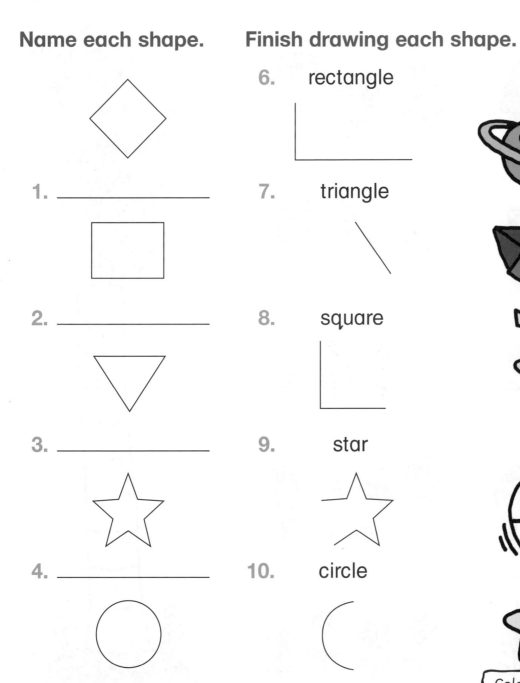

1. _____

2. _____

3. _____

4. _____

5.

6. rectangle

7. triangle

8. square

9. star

10. circle

Colour in your score

36

Test 6 Counting sequences to 12

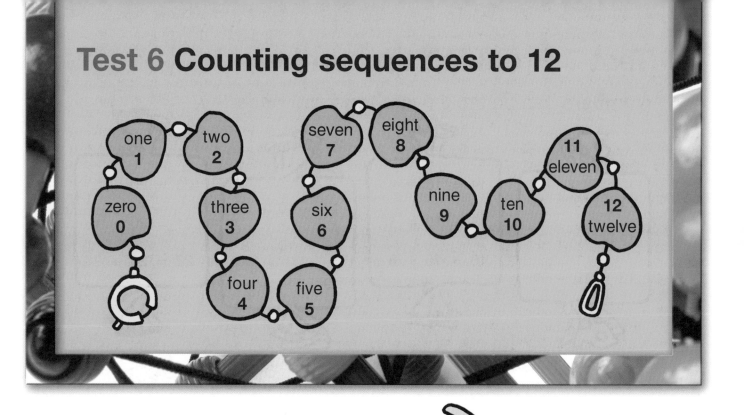

Write the missing numbers.

1. | 2 | 3 | 4 | | |

6. | | | 10 | 11 | 12 |

2. | | | 7 | 8 | 9 |

7. | | 4 | 5 | | |

3. | | | 9 | 10 | |

8. | 12 | 11 | 10 | | |

4. | 9 | 8 | 7 | | |

9. | | | 6 | 5 | 4 |

5. | | | 4 | 3 | |

10. | | | | 8 | 7 |

Colour in your score

37

Test 7 Breaking up numbers

Numbers can be broken into **tens** and **ones**.

11 = 10 + 1

12 = 10 + 2

13 = 10 + 3

14 = 10 + 4

15 = 10 + 5

16 = 10 + 6

17 = 10 + 7

18 = 10 + 8

19 = 10 + 9

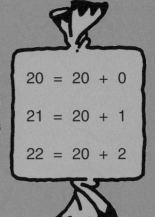

20 = 20 + 0

21 = 20 + 1

22 = 20 + 2

Fill in the missing numbers.

1. 16 = 10 + ☐

2. 14 = 10 + ☐

3. 17 = 10 + ☐

4. 18 = 10 + ☐

5. 21 = 20 + ☐

6. 17 = ☐ + 7

7. 13 = ☐ + 3

8. 12 = ☐ + 2

9. 19 = ☐ + 9

10. 11 = ☐ + 1

Colour in your score

38

Test 8 Adding and subtracting to 20

When **adding**, **count on** from the **larger number**.

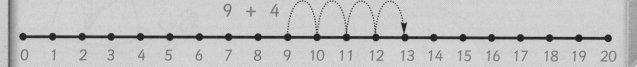

$$9 + 4$$

0 1 2 3 4 5 6 7 8 9 10 11 12 13 14 15 16 17 18 19 20

When **subtracting**, **count back** from the **larger number**.

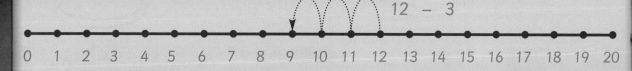

$$12 - 3$$

0 1 2 3 4 5 6 7 8 9 10 11 12 13 14 15 16 17 18 19 20

Write the answers.

1. $8 + 5 =$

2. $2 + 13 =$

3. $11 + 4 =$

4. $3 + 17 =$

5. $15 + 4 =$

6. $16 - 3 =$

7. $20 - 2 =$

8. $18 - 5 =$

9. $11 - 2 =$

10. $15 - 4 =$

Colour in your score

10 · 9 · 8 · 7 · 6 · 5 · 4 · 3 · 2 · 1

Test 9 Time: o'clock

The **minute hand** points to **12** for **o'clock times**.

8 o'clock

Write the times.

1.  ☐ o'clock

2.  ☐ o'clock

3.  ☐ o'clock

4.  ☐ o'clock

5.  ☐ o'clock

Draw the missing hand.

6.

six o'clock

7.

four o'clock

8.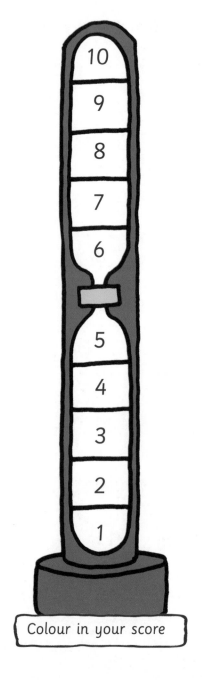

eight o'clock

9.

ten o'clock

10.

two o'clock

Colour in your score

40

Test 10 Data diagrams

Some diagrams show **opposites**.

square	not square

Draw each shape on the diagram.

1.	2.	3.	4.	5.

straight sides	not straight sides

Write each number on the diagram.

6. 3   7. 11   8. 15   9. 8   10. 10

less than 10	not less than 10

Colour in your score

41

Say these numbers **forwards** and **backwards**.

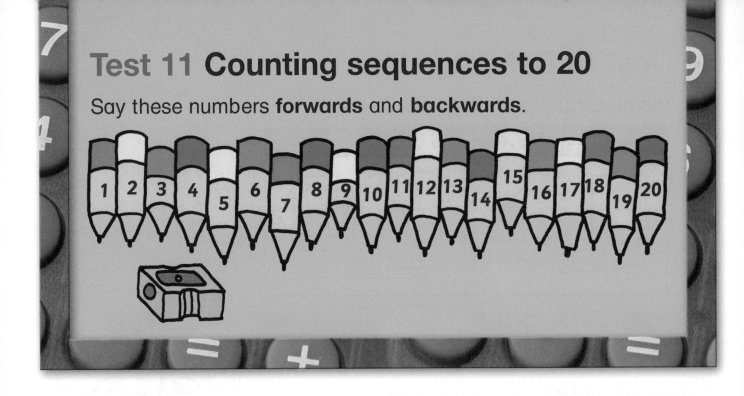

Write the missing numbers.

1.
8 9 10

6.
14 15 16

2.
10 11 12

7.
8 9 10

3.
15 16 17

8.
20 19 18

4.
15 14 13

9.
16 15 14

5.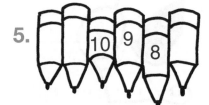
10 9 8

10.
19   17   15

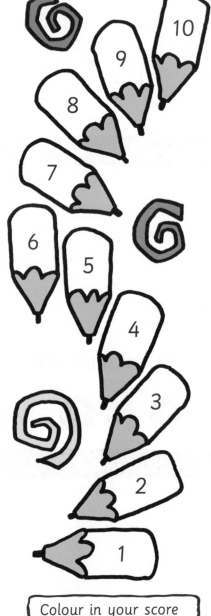

Colour in your score

Test 12 Ordinal numbers

Some numbers show the **order** of things.

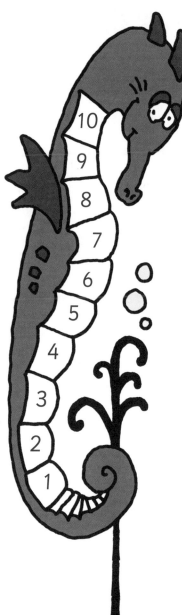

1. Colour the 3rd snail.

2. Colour the 1st star.

3. Colour the 4th spider.

4. Colour the 2nd ladybird.

5. Colour the last worm.

6. Colour the 3rd letter.

PETER

7. Colour the 4th letter.

PAUL

8. Colour the 5th letter.

SALLY

9. Colour the 2nd letter.

HARRY

10. Colour the last letter.

GITA

Colour in your score

43

Test 13 Subtraction within 10

Jumping **back** is the same as **subtraction**.

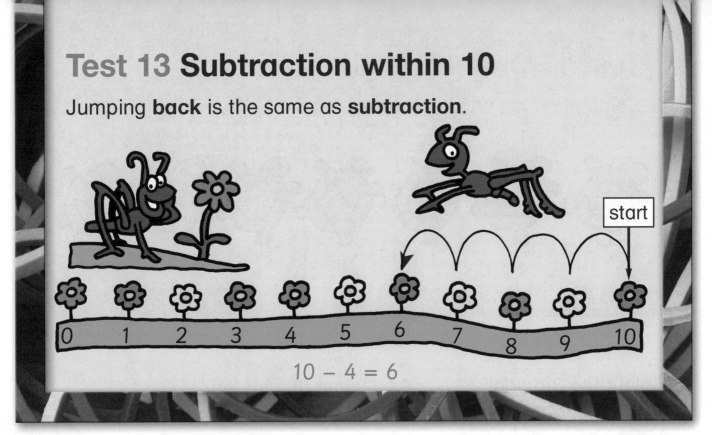

$$10 - 4 = 6$$

Use the number line to help answer these.

1. 7 – 2 =

2. 8 – 4 =

3. 5 – 5 =

4. 9 – 1 =

5. 8 – 6 =

6. 7 – 5 =

7. 10 – 8 =

8. 6 – 4 =

9. 10 – 5 =

10. 9 – 3 =

Colour in your score

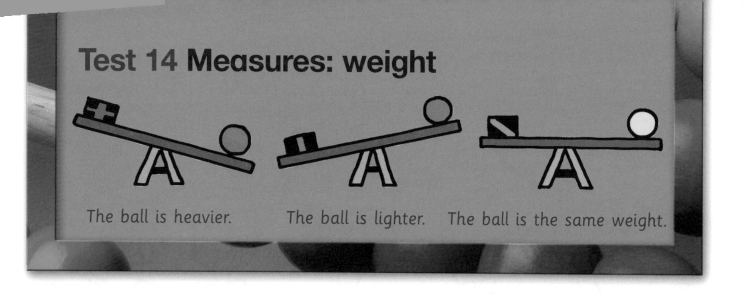

Test 14 Measures: weight

The ball is heavier.   The ball is lighter.   The ball is the same weight.

Underline the word to show whether the ball is heavier, lighter or the same weight.

1. heavier lighter same

2. heavier lighter same

3. heavier lighter same

4. heavier lighter same

5. heavier lighter same

6. heavier lighter same

7. heavier lighter same

8. heavier lighter same

9. heavier lighter same

10. heavier lighter same

Colour in your score

Test 15 3-D shapes

Here are the **names** of some **shapes**.

sphere   cube   cuboid   cylinder   cone

Join each shape to its name.

1.    sphere   6.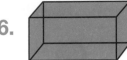

2.    cone   7.

3.   cuboid   8.

4.    cylinder   9.

5. 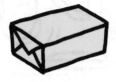   cube   10.

Colour in your score

Test 16 Counting on and back

It is very useful to be able to **count on** and **back**.

```
0  1  2  3  4  5  6  7  8  9  10  11  12  13  14  15  16  17  18  19  20
```

count on 3   count back 3

Write these missing numbers.

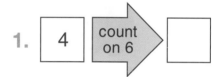
1. 4 | count on 6 | ☐

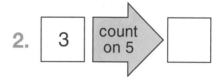
6. count on 7 ☐ 12

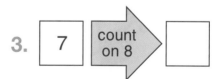
2. 3 | count on 5 | ☐

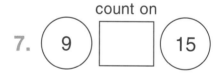
7. count on 9 ☐ 15

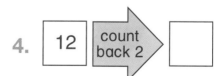
3. 7 | count on 8 | ☐

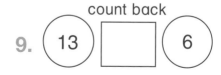
8. count back 12 ☐ 8

4. 12 | count back 2 | ☐

9. count back 13 ☐ 6

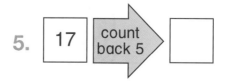
5. 17 | count back 5 | ☐

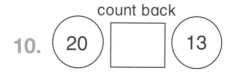
10. count back 20 ☐ 13

10 9 8 7 6 5 4 3 2 1

Colour in your score

47

Test 17 Comparing numbers

13 is bigger than **9**   **6** is smaller than **12**

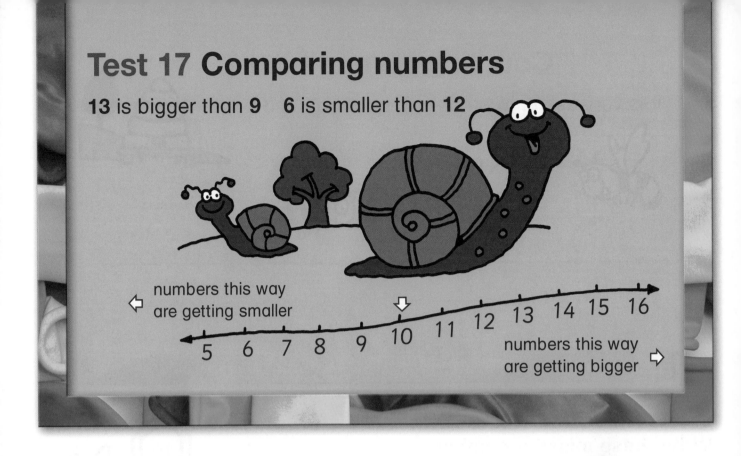

numbers this way are getting smaller

5 6 7 8 9 10 11 12 13 14 15 16

numbers this way are getting bigger

Colour the bigger number.

1.  12 7

2.  11 14

3.  15 12

4.  10 11

5.  13 16

Colour the smaller number.

6.  4 7

7.  11 19

8.  13 12

9.  15 8

10.  13 17

10 9 8 7 6 5 4 3 2 1

Colour in your score

48

Test 18 Addition using doubles

Near doubles can help us add.

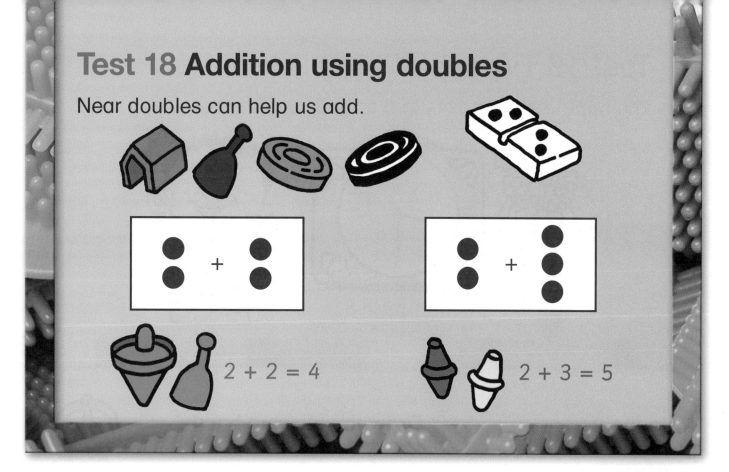

2 + 2 = 4

2 + 3 = 5

Add these doubles.

1. ● ● = ☐

2. [dice] [dice] = ☐

3. [dice] [dice] = ☐

4. [dice] [dice] = ☐

5. [dice] [dice] = ☐

Add these near doubles.

6. 4 3 = ☐

7. 2 3 = ☐

8. 5 6 = ☐

9. 4 3 = ☐

10. 4 5 = ☐

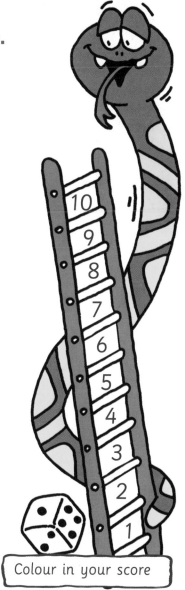

Colour in your score

49

Test 19 Time: half past

The **minute hand** points to **6** for **half past** times.

half past 8

Write the times.

1. half past ☐

2. half past ☐

3. half past ☐

4. half past ☐

5. half past ☐

Draw the missing hand.

6. half past 1

7. half past 11

8. half past 4

9. half past 7

10. half past 9

Colour in your score

50

Test 20 Data graphs

This **graph** shows children's favourite ice-creams.

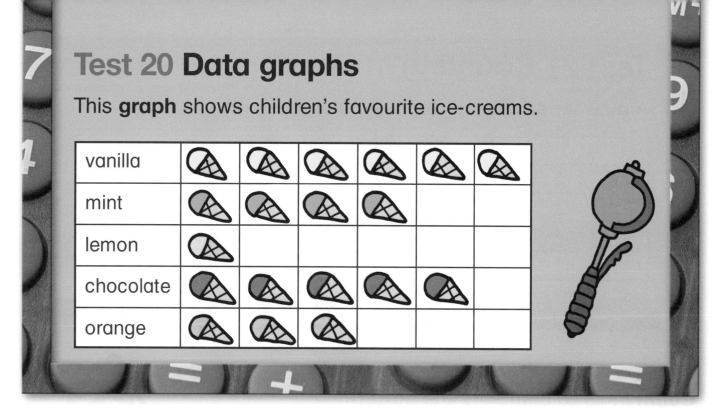

vanilla						
mint						
lemon						
chocolate						
orange						

How many children's favourite was:

1. chocolate?

2. mint?

3. vanilla?

4. orange?

5. lemon?

6. Which was the favourite?

7. Which was the least favourite?

8. Which was chosen by 4 children?

9. Which was chosen by 3 children?

10. How many children were there altogether?

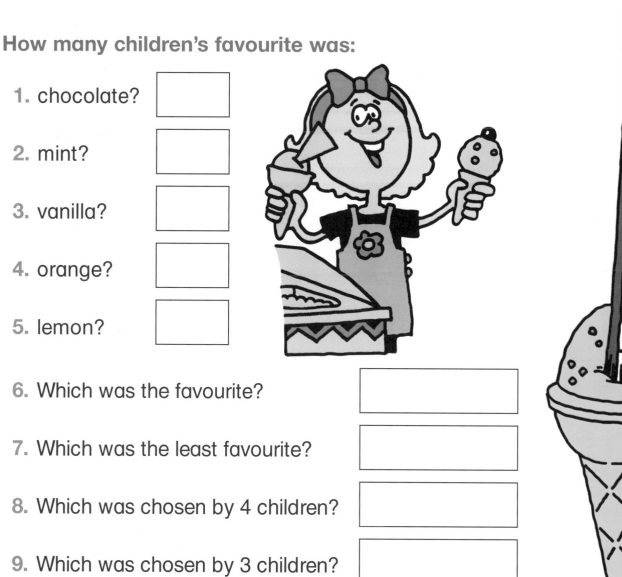

Colour in your score

Test 21 Counting in steps

A **number grid** helps you to see **number patterns**.

1	2	3	4	5	6	7	8	9	10
11	12	13	14	15	16	17	18	19	20
21	22	23	24	25	26	27	28	29	30
31	32	33	34	35	36	37	38	39	40
41	42	43	44	45	46	47	48	49	50

Count in 2s to complete the pattern.

1. | 4 | 6 | 8 | | | |

2. | 30 | 28 | 26 | | | |

3. | 3 | 5 | 7 | | | |

4. | 15 | 17 | 19 | | | |

5. | 21 | 19 | 17 | | | |

Count in 5s to complete the pattern.

6. | 5 | 10 | 15 | | | |

7. | 50 | 45 | 40 | | | |

8. | 1 | 6 | 11 | | | |

9. | 2 | 7 | 12 | | | |

10. | 30 | 25 | 20 | | | |

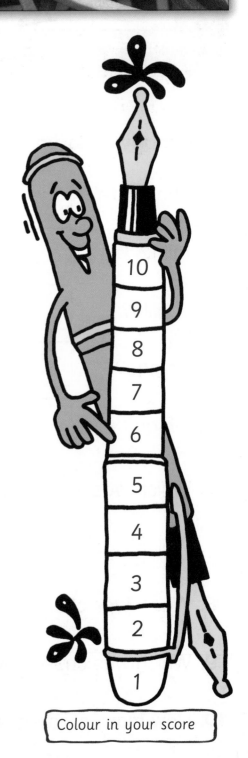

Colour in your score

Test 22 More or less

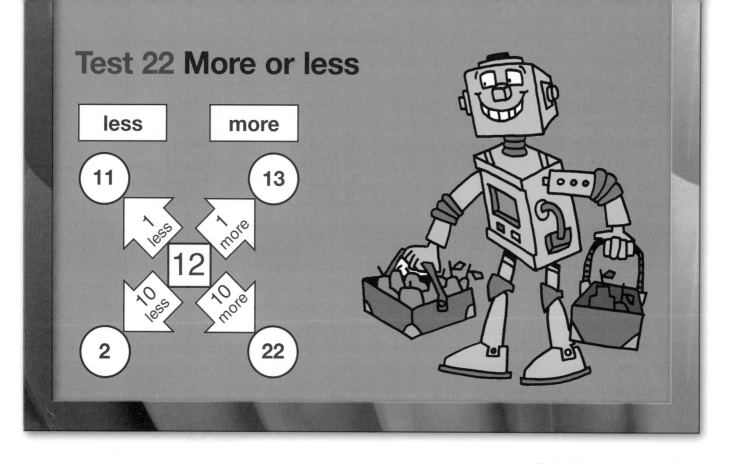

less   more

11   13

1 less   1 more

12

10 less   10 more

2   22

Make less.

1. 17 [1 less] ◯

2. 21 [1 less] ◯

3. 25 [10 less] ◯

4. 31 [10 less] ◯

5. 35 [10 less] ◯

Make more.

6. 5 [1 more] ◯

7. 8 [1 more] ◯

8. 6 [10 more] ◯

9. 11 [10 more] ◯

10. 15 [10 more] ◯

10
9
8
7
6
5
4
3
2
1

Colour in your score

53

Test 23 Adding and subtracting

A **number line** will help with **adding** and **subtracting**.

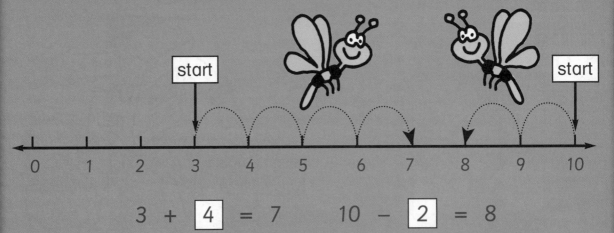

$$3 + \boxed{4} = 7 \qquad 10 - \boxed{2} = 8$$

Write the missing number in the box.

1. $3 + \boxed{} = 10$

2. $2 + \boxed{} = 7$

3. $7 + \boxed{} = 10$

4. $4 + \boxed{} = 8$

5. $3 + \boxed{} = 9$

6. $8 - \boxed{} = 4$

7. $7 - \boxed{} = 1$

8. $5 - \boxed{} = 2$

9. $9 - \boxed{} = 6$

10. $4 - \boxed{} = 0$

Colour in your score

Test 24 Measures: capacity

full   nearly full   half full   nearly empty   empty

Colour the mugs and bowls to show how full they are.

1. half full

2. empty

3. nearly full

4. nearly empty

5. full

6. full

7. nearly empty

8. empty

9. nearly full

10. half full

Colour in your score

Test 25 Shapes patterns

Look at these **patterns**.

/ / O / / O / / O	△ □ △ □ △ □ △ □
2 2 3 2 2 3 2 2 3	∨ ∩ ∨ ∩ ∨ ∩ ∨ ∩ ∨ ∩

Continue each pattern.

1. ∧∧∧

2. (wave pattern)

3. ⊓⊔⊓ (square wave pattern)

4. (curved notch pattern)

5. ∧∧

6. 1 1 2 1 1 2

7. |o|o|o|

8. □□•□□•

9. △□∘△□∘

10. 1 2 3 1 2 3

10
9
8
7
6
5
4
3
2
1

Colour in your score

Test 26 Odds and evens

even numbers						odd numbers				
2	4	6	8	10		1	3	5	7	9
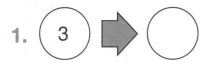						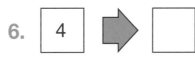				
These end in 2, 4, 6, 8, or 0.						These end in 1, 3, 5, 7, or 9.				

Write the next odd number.

1. (3) ➡ ( )

2. (7) ➡ ( )

3. (11) ➡ ( )

4. (15) ➡ ( )

5. (19) ➡ ( )

Write the next even number.

6. [4] ➡ [ ]

7. [8] ➡ [ ]

8. [12] ➡ [ ]

9. [16] ➡ [ ]

10. [20] ➡ [ ]

Colour in your score

Number tracks help us to put numbers in **order**.

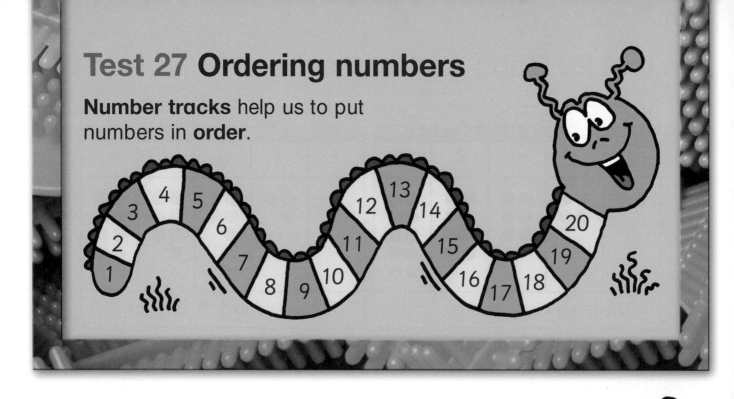

Write these numbers in order.

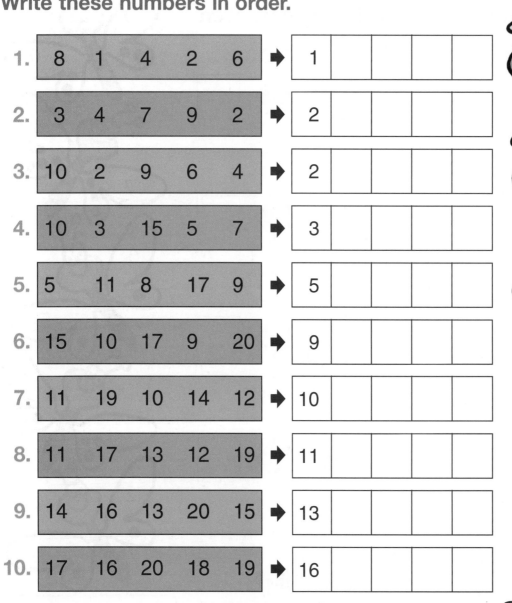

1. 8 1 4 2 6 ➡ 1 _ _ _ _

2. 3 4 7 9 2 ➡ 2 _ _ _ _

3. 10 2 9 6 4 ➡ 2 _ _ _ _

4. 10 3 15 5 7 ➡ 3 _ _ _ _

5. 5 11 8 17 9 ➡ 5 _ _ _ _

6. 15 10 17 9 20 ➡ 9 _ _ _ _

7. 11 19 10 14 12 ➡ 10 _ _ _ _

8. 11 17 13 12 19 ➡ 11 _ _ _ _

9. 14 16 13 20 15 ➡ 13 _ _ _ _

10. 17 16 20 18 19 ➡ 16 _ _ _ _

Colour in your score

58

Test 28 Addition patterns

Addition facts to **10** are important.

10 + 0 0 + 10	9 + 1 1 + 9	8 + 2 2 + 8	7 + 3 3 + 7	6 + 4 4 + 6	5 + 5

10

Write the missing number in the box.

1.   7 + ☐ = 10

2.   4 + ☐ = 10

3.   8 + ☐ = 10

4.   3 + ☐ = 10

5.   9 + ☐ = 10

6.   ☐ + 1 = 10

7.   ☐ + 4 = 10

8.   ☐ + 6 = 10

9.   ☐ + 8 = 10

10. ☐ + 3 = 10

Colour in your score

Test 29 Days of the week

Learn the **order** of the **days**.

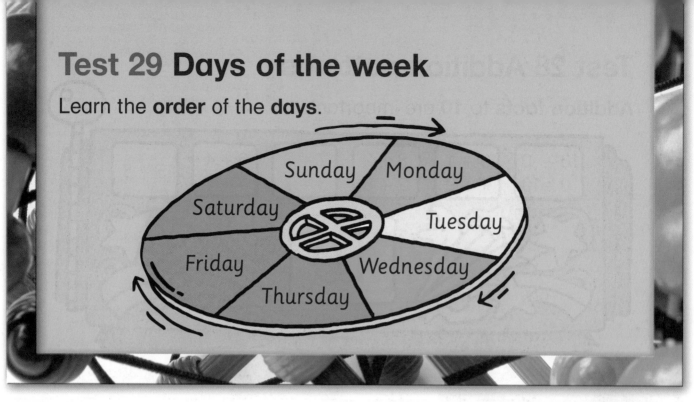

Write the day before:

1. Monday

2. Thursday

3. Wednesday

4. Tuesday

5. Saturday

Write the day after:

6. Friday

7. Sunday

8. Tuesday

9. Saturday

10. Wednesday

Colour in your score

Test 30 Learning to tally

Tally marks show how **many**.

1	2	3	4	5
|	||	|||	||||	HHH

6	7	8	9	10
HHH |	HHH ||	HHH |||	HHH ||||	HHH HHH

Make tally marks to show how many fish and insects there are.

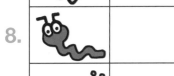

1.

2.

3.

4.

5.

6.

7.

8.

9.

10.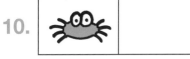

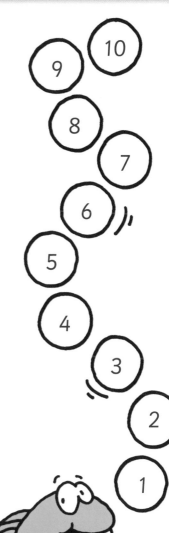

Colour in your score

ANSWERS

Page 2

I

II
one → 1	six → 6
two → 2	seven → 7
three → 3	eight → 8
four → 4	nine → 9
five → 5	ten → 10

Page 3

I **a** 8   **b** 6   **c** 9
d 11   **e** 14   **f** 15

II There are 20 fish altogether.

Page 4

I **a** 3 and 2 makes a total of 5

b 5 and 3 makes a total of 8

c 4 and 3 makes a total of 7

II Check there are now 8 beans in each set.

Page 5

I
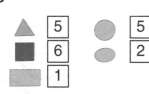

II
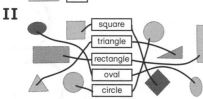

square
triangle
rectangle
oval
circle

Page 6

I a

b

II a 3 pennies
b 5 pennies
c 2 pennies
d 4 pennies
e 3 pennies

Page 7

I a 7 o'clock
b 1 o'clock
c 6 o'clock
d 8 o'clock
e 9 o'clock
f 3 o'clock
g 11 o'clock
h 10 o'clock

II a   **c**

b   **d**

Page 8

I Check child's writing.

II a 14   **b** 18   **c** 13
d 16   **e** 19   **f** 12
g 11   **h** 15

Page 9

I Check child's patterns.

II Check child's patterns.

Page 10

I a 7, 8, 9, 10, 11, 12, 13, 14, 15, 16

b 11, 12, 13, 14, 15, 16, 17, 18, 19, 20

c 18, 17, 16, 15, 14, 13, 12, 11, 10, 9

d 12, 11, 10, 9, 8, 7, 6, 5, 4, 3

II

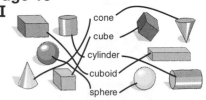

The shaded number is 8.

Page 11

I a 5   **d** 14   **g** 13
b 10   **e** 6   **h** 9
c 16   **f** 8

II a 3 4 5 6 ⑦ 8 9
b 5 6 7 8 ⑨ 10 11
c 2 3 4 ⑤ 6 7 8
d 6 ⑦ 8 9 10 11 12

Page 12

I a 3 + 2 = 5   **c** 4 + 4 = 8
b 3 + 3 = 6   **d** 5 + 3 = 8

II a 4   **c** 2
b 3   **d** 4

Page 13

I

cone
cube
cylinder
cuboid
sphere

II Check child has coloured a, g and h.

Page 14

I Check child has crossed out correct number of buns.
a 1   **c** 2
b 4   **d** 3

II a 6 − 4 = 2   **c** 5 − 3 = 2
b 7 − 3 = 4   **d** 6 − 5 = 1

Page 15

I Check these coins have been crossed out.
a 1p   **d** 20p
b 2p   **e** 50p
c 10p   **f** 1p

II a 5p   **c** 14p
  b 8p   **d** 13p

Page 16
I a 13 → 14   18 → 19
     26 → 27   31 → 32
  b 19 → 18   15 → 14
     26 → 25   20 → 19
  c 7 → 17   19 → 29
     14 → 24   22 → 32
  d 18 → 8   21 → 11
     32 → 22   29 → 19

II a 5p   16p
  b 16p   30p

Page 17
I a 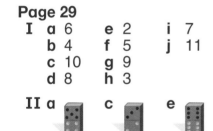 **b** (shields: 12 18 14 / 11 3 16 / 12 2 20)

II a 11→13, 15→17, 13→15
  b 8→10, 16→18, 20→22

Page 18
I a 7, 8, 9, 10, 11, 12, 13
  b 14, 15, 16, 17, 18, 19, 20
  c 19, 18, 17, 16, 15, 14, 13
  d 15, 14, 13, 12, 11, 10, 9

II Check cans are joined in this order: 7p, 9p, 11p, 13p, 14p, 15p, 18p, 20p

Page 19
I a 10 + 5   **f** 10 + 2
  b 10 + 8   **g** 10 + 4
  c 10 + 1   **h** 10 + 3
  d 10 + 6   **i** 10 + 7
  e 10 + 9   **j** 10 + 10

II a thirteen   **e** twelve
  b eleven   **f** sixteen
  c nineteen   **g** seventeen
  d fourteen   **h** eighteen

Page 20
I a (6 9 / 13 14 / 21 19) **b** (11 7 / 18 19 / 21 23) (9 13 / 14 10 / 18 25)

II a 6, 7, 8, 9
  b 11, 12, 13, 14
  c 17, 18, 19, 20
  d 12, 13, 14, 15
  e 19, 20, 21, 22
  f 23, 24, 25, 26

Page 21
I The 1st letter is A.
The 5th letter is E.
The 7th letter is G.
The 4th letter is D.
The last letter is Z.
F is the 6th letter.
C is the 3rd letter.
L is the 12th letter.
P is the 16th letter.
B is the 2nd letter.

II a MATHS
  b MAGIC

Page 22
I a 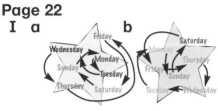 **b**

II Wednesday   Sunday
Saturday   Thursday
Friday   Tuesday
Monday

Page 23
I a 12p   **c** 8p
  b 30p   **d** 18p

II a (coins)   **c** (coins)

  b (coins)

Page 24
I a 7   **f** 7   **k** 9
  b 4   **g** 8   **l** 9
  c 6   **h** 5   **m** 6
  d 2   **i** 5   **n** 10
  e 6   **j** 8   **o** 10

II a 7   **d** 10   **g** 9
  b 9   **e** 7   **h** 6
  c 6   **f** 10

Page 25
I a 5   **c** 4   **e** 3
  b 4   **d** 3   **f** 2

II a odd one out 5 − 2
  b odd one out 4 − 0
  c odd one out 6 − 5

Page 26
I a 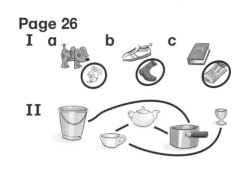 **b**   **c**

II

Page 27
I The numbers 2, 4, 6, 8, 10, 12, 14, 16, 18 and 20 should be coloured red.
  a even
  b odd

II a 16, 15, 14, 13
  b 12, 14, 16, 18
  c 17, 19, 21, 23
  d 25, 30, 35, 40
  e 22, 20, 18, 16

Page 28
I a 2   **c** 4   **e** 8
  b 10   **d** 5   **f** 3

II a   **b**   **c**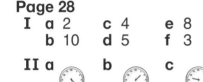

Page 29
I a 6   **e** 2   **i** 7
  b 4   **f** 5   **j** 11
  c 10   **g** 9
  d 8   **h** 3

II a   **c**   **e**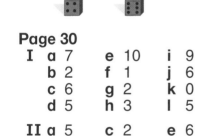
  b   **d**

Page 30
I a 7   **e** 10   **i** 9
  b 2   **f** 1   **j** 6
  c 6   **g** 2   **k** 0
  d 5   **h** 3   **l** 5

II a 5   **c** 2   **e** 6
  b 3   **d** 4

Page 31
I a 4   **c** 3
  b 4   **d** 5

II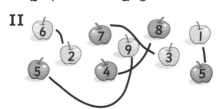

Page 32
1.	1	6.	7
2.	5	7.	9
3.	2	8.	4
4.	8	9.	7
5.	3	10.	10

Page 33
1.	6	6.	three
2.	8	7.	nine
3.	12	8.	eleven
4.	15	9.	seventeen
5.	20	10.	nineteen

Page 34
1.	9	6.	8
2.	7	7.	8
3.	9	8.	10
4.	8	9.	10
5.	9	10.	9

Page 35
1. ☐ ☑ ☐
2. ☐ ☐ ☑
3. ☑ ☐ ☐
4. ☐ ☑ ☐ ☐
5. ☐ ☐ ☐ ☑
6. ☐ ☑ ☐
7. ☑ ☐ ☐
8. ☐ ☐ ☐ ☑
9. ☐ ☑ ☐ ☐
10. ☐ ☐ ☐ ☑

Page 36
1. square
2. rectangle
3. triangle
4. star
5. circle

6. ▭
7. △
8. ▭
9. ☆
10. ◯

Page 37
The missing numbers are in **bold**.
1. 2 3 4 **5 6**
2. **5 6** 7 8 9
3. **7 8** 9 10 **11**
4. 9 8 7 **6 5**
5. **6 5** 4 3 **2**
6. **8 9** 10 11 12
7. **3** 4 5 **6 7**
8. 12 11 10 **9 8**
9. **8** 7 6 5 4
10. **10 9** 8 7 **6**

Page 38
1.	6	6.	10
2.	4	7.	10
3.	7	8.	10
4.	8	9.	10
5.	1	10.	10

Page 39
1.	13	6.	13
2.	15	7.	18
3.	15	8.	13
4.	20	9.	9
5.	19	10.	11

Page 40
1. 3 o'clock
2. 9 o'clock
3. 1 o'clock
4. 10 o'clock
5. 5 o'clock

6. 🕕 9. 🕙
7. 🕑 10. 🕑
8. 🕚

Page 41
1. straight sides
2. not straight sides
3. straight sides
4. not straight sides
5. straight sides
6. less than 10
7. not less than 10
8. not less than 10
9. less than 10
10. not less than 10

Page 42
The missing numbers are in **bold**.
1. 8 9 10 **11 12 13**
2. **8 9** 10 11 12 **13**
3. **14** 15 16 17 **18 19**
4. 15 14 13 **12 11 10**
5. **12 11** 10 9 8 7
6. 14 15 16 **17 18 19**
7. 7 8 9 10 **11 12**
8. 20 19 18 **17 16 15**
9. **17** 16 15 14 **13 12**
10. 19 **18** 17 **16** 15 **14**

Page 43
1. 🐱🐱🐱🐱🐱
2. ⭐☆⭐☆⭐
3. 🕷🕷🕷🕷
4. 🐞🐞🐞🐞🐞
5. 🐛🐛🐛🐛
6. T
7. L
8. Y
9. A
10. A

Page 44
1.	5	6.	2
2.	4	7.	2
3.	0	8.	2
4.	8	9.	5
5.	2	10.	6

Page 45
1.	same	6.	same
2.	heavier	7.	lighter
3.	heavier	8.	heavier
4.	lighter	9.	same
5.	lighter	10.	lighter

Page 46
1.	sphere	6.	cuboid
2.	cube	7.	cone
3.	cylinder	8.	cube
4.	cone	9.	sphere
5.	cuboid	10.	cylinder

Page 47
1.	10	6.	5
2.	8	7.	6
3.	15	8.	4
4.	10	9.	7
5.	12	10.	7